# COLORING BOOK
## WONDER WORLDS 2

## BY BELLA STITT

# INTRODUCTION

Certified cognitive therapist Bella Stitt developed this coloring book for relieving stress from everyday life. Each picture is a different wonder world, appearance of which depends on the way you choose to color it.

As a result of coloring of these pictures, you will feel less tense and more relaxed. After you are finished coloring, you can frame your art as a reminder of your accomplishment and your ability to release stress and better manage your physical and mental state of mind.

Feel free to color as you listen to music or watch television. Coloring will put you at ease and make you better able to focus. This is a form of meditation as well as an opportunity to think about your struggles and come up with solutions while engaging in a calming activity.

In addition, if you have difficulty with perfectionism and making decisions, choosing colors to use for these pictures will allow you to have practice with experiencing flexibility and courage while having a positive outcome since there is no right or wrong way to color.

Relax and enjoy!